2 Beginners Edition
Ebonygeek45

s o s

. . . - - - . . .

Project 2

Uno Programming

digitalWrite

More On C++

Shannon Davis

Ebonynerd45@gmail.com

ISBN-10: 1543008380

ISBN-13: 978-1543008388

DEDICATION

When ever I need someone to support my efforts. Who have as much faith in projects and plans as much as I do. She always have my back. I can always count of my baby sister Latanya Davis. I am so lucky she is in my life.

Ebonygeek aka Shannon Davis

Also By Shannon Davis/Ebonygeek45

Books

Uno Programming digitalWrite: Project 1 S O S

Beginners Edition 1

Videos

Ebonygeek45: Youtube Channel

CONTENTS

INTRODUCTIONS

The programming, electronics, prototyping, and the world in general is huge. People live, work, and play in this world. We love the new gadgets and gizmos that come out.

Life is good isn't it?

Children get these toys as a gift or just begged for it until they got it.

That is when the love for new exciting things start. They play with it, bang it around, uhh oh it is broken.

That is the life of gadgets and gizmos. From childhood to retirement and beyond we love technology.

Some kids ask why?

So they find a screw driver or in my childhood a butter knife, or even pull it apart. They pull out the stuffing or break apart any casings and find electronics. Some kids toy with it and get a beep or movement.

The minds of the curious is tweaked.

They ask why?

Most of the time they get a lecture or even a spanking or time out.

How could you break that new(and most of the time expensive) toy?

But sometimes, more rarely then most an adult or even another kid can answer to them – WHY.

Those fortunate kids grow up and keep tinkering and learning.

Some adults go through the same process later in life. They are called geeks and nerds and picked on like that is a negative thing. It is not and most of them keep expanding their knowledge and grow.

They are the troubleshooters, problem solvers, and go to people of the world.

The workforce of the world reward and acknowledge these fortunate ones. These ones are rarer then most.

For the larger part of the workforce theses special ones are

underpaid, overworked, demoralized, and expendable.

Maybe even further used and exploited by student lending in the business of college education for a profit.

Opportunist, exploiters. leeches, users, spongers, parasites, social

engineers, etc all come to mind in the real work world.

But they tinker on and that is the strength of these people. They are

inventors, engineers, and genius of the day.

Nikola Tesla, Micheal Faraday, Thomas Edison, Benjamin Franklin,

Marie Curie, Charles Babbage, Robert Kahn, Vint Cerf, to name a

few. Later on you have the notable of the day like: Bill Gates, Linus Torvalds, Massimo Banzi, Robert Portugal, comes to mind.

There are countless others. The nameless and faceless ones that we are not aware of. But they are just as notable.

Then you have your common everyday men and women.

They tinker on in basements, garages, at home in laboratory's, workshops, incubators, etc.

They find materials at junk yards, stores, and on the internet.

Toys, computers, and electronics taken apart and reused to make incredible things. They are anywhere from the young to the old. Spanning through the wealthy to the impoverished.

They do it for there own personal fulfillment or gain. To fix something that is broken. To make something for someone as a gift. Or any other small reason they can find. To sell or be the first to come out with an idea people can improve on.

These people are the salt of the earth and the reason we enjoy what we enjoy today.

We need to encourage this. Catch people like this when they are young for special education and training.

Encourage youths and adults to understand and create the technology they so love.

Wouldn't that be a wonderful world?

Imagine how much further we would be.

It has already started. Through Wikipedia, YouTube, Github, and open sourcing. More need to be done to get us there. How fast may very well depend on you and me.

1 CHAPTER COMMUNICATE WITH YOUR UNO

Congratulations on getting through the 1st edition of this series. If you did not read that book, you may want to go back to it. It has a lot of good information. This book is starting where that one left off.

It is:

Uno Programming digitalWrite: Project 1 S O S

Beginners Edition 1

My apologies for the Table of Contents in that book. It came to my attention that somehow it didn't come out well.

To avoid that problem with this book, The chapters and sub chapters are not number.

The best way to work with that is to copy whatever chapter or sub chapter you are looking for and use find to find it.

But, let's get to what we are here for.

Getting Started

To refresh we left off with:

The pin out below

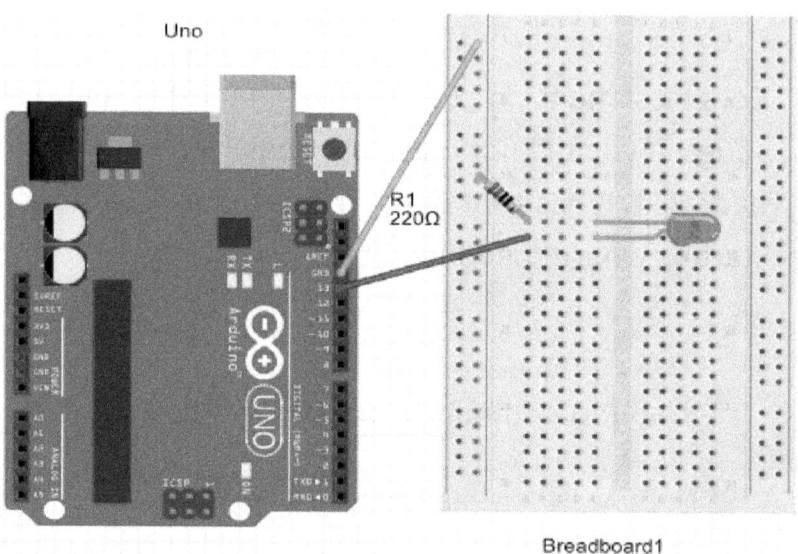

Uno

R1
220Ω

Breadboard1

The sketch below.

```
// morseSos.ino

// Attributes

const int pin13 = 13;      // Stores pin number

const int dwDot = 300;   // Stores 300 millisec value

const int dwDash = 900;  // Stores 900 millisec value

const int dwBetween = 1000;  // Stores 1 sec value
```

```
/*

 setup function runs once when you press reset or

power the board

*/

void setup()

{

    // initialize digital pin 13 as an output.

    pinMode(pin13, OUTPUT);

}

// the loop function runs over and over again forever

void loop()

{

    /// S

    // .

    theBlinks(dwDot, pin13, dwBetween);
```

```
/// O

// -

theBlinks(dwDash, pin13, dwBetween);



/// S

// .

theBlinks(dwDot, pin13, dwBetween);



}



// Behaviors


int theBlinks(int passing1, int passing2, int passing3)

{
```

```
for (int e = 0; e < 3; e++)

{

    digitalWrite(pin13, HIGH);

    delay(passing1);

    digitalWrite(pin13, LOW);

    delay(passing1);

}


// delay 1 seconds between letters

delay(passing3);        // wait for a second

}
```

If you are anything like me I know you have came up
with some. interesting projects of your own.

Just so we are on the same page it is a good idea to use

the sketch and the circuit above.

Our program blinks the sos pattern.

Wouldn't it be nice for it to also tell us what it is blinking in letters?

Believe it or not we can program it to do just that.

That is one of the good things about using a micro — controller, in other words the Uno. It is a small computer which is the biggest advantage.

However you do have to program it to communicate with you correctly. That is where our C++ comes in again.

Again the good Arduino IDE folks helps to make it simple for you to achieve this.

The Serial Monitor

Remember the Serial Monitor Icon we went over before.

Underneath the X in the upper corner of the Arduino is the icon for your Serial Monitor as pictured .

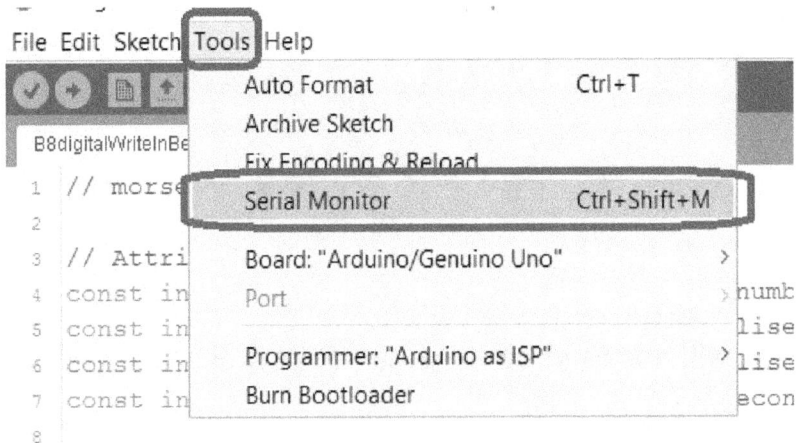

You can also go to Tools > Serial Monitor

Both ways will bring up your Serial Monitor.

Remember to have your Uno plugged in. If you don't you will get an error because it don't have anything to communicate with.

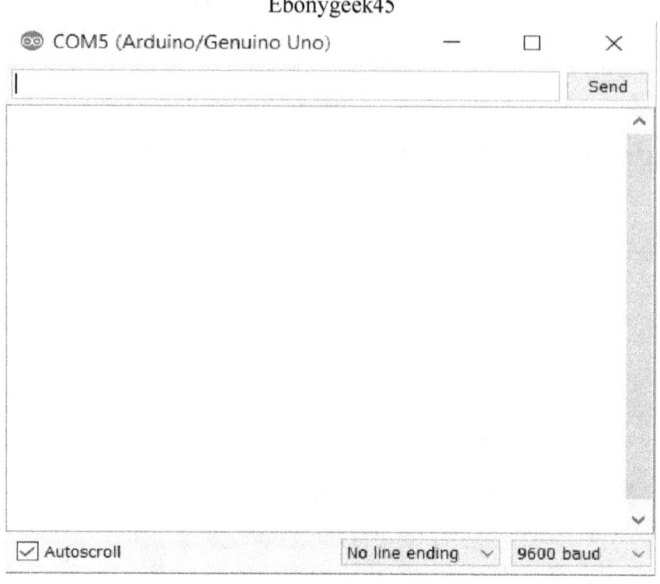

It will not be showing data yet because we have not program it to.

The serial monitor can be used for a great many things. It can be used to show information coming in from a sensor for example. You can also send data to a component. These are topics and projects that will be down the line.

One of my favorite uses for the serial monitor is for troubleshooting my code. This can be extremely helpful down the line. Because it allows you to kind of step through your code.

Printing To Serial Monitor

For now we want it to print to the screen to show what our blinks are doing.

The syntax for it is:

For the setup function. Code that run once:

Serial.begin(<baud>);

What is baud? You may be asking.

Baud sets data rate by bits for serial transmission. In other words it is the speed the data flows. The default is 9600 in most cases.

For the loop funtion

Serial.print(<variable>);

The print allow the variable to print on the same line.

or

Serial.println(<variable>);

... println will go to the next line after it has printed what you set up.

There are a lot of other functions to use with the serial monitor already made out for you. Check their site for more www.arduino.cc .

For now the above syntax is all we need for what we are doing.

Go ahead and close out the serial monitor.

An Algorithm Coming On

Ok I am not going to keep going on about algorithms in this book. I stress to use them.

Here is the one we will have for the code we are going to be doing

1. Add attributes(variables) for what we will need done

2. Use those attributes(variables) where needed in our behaviors(functions).

3. Go into our behaviors(functions) and add the code needed to print to the serial monitor.

4. Verify and upload

5. Troubleshoot as needed

6. Have a grape(much healthier than a cookie).

Take a good look at the code and see how to work through the Algorithm.

See if you can work it out by yourself. If you can great. You will not be doing anything that was not explained:

Uno Programming digitalWrite: Project 1 S O S

Beginners Edition 1

If you can't figure it come back to this point and we will go through it.

You need to work through it and put on your thinking caps.

Working Through The Algorithm

If you have worked through the algorithm and checked the serial monitor to see if it is printing to the screen correctly, great. If it is also blinking correctly, good job.

Your code may not look like the solution given. That does not mean your code is more wrong or more right. There are many ways to program.

Now we will add the serial monitor to the mix, and work through the algorithm.

1. Add attributes for what we will need done

Attributes are the variables needed. So we add that with the other attributes we already have.

```
3   // Attributes
4   const int pin13 = 13;          // Stores pin number
5   const int dwDot = 300;      // Stores 300 millisecond value
6   const int dwDash = 900;    // Stores 900 millisecond value
7   const int dwBetween = 1000;    // Stores 1 second value
8
9   /* 1 Set up Attributes
10   *    For . and - Then S and O.
11   */
12  // Attributes to print to serial monitor
13  const char dwDotPrintSym = '.';
14  const char dwDashPrintSym = '-';
15  const char dwDotPrintLet = 'S', dwDashPrintLet = 'O';
```

Instead of using int we are using the char type. When using the char type you would use single quotes. If you try to use double quotes for the char type you will get an error. The double quotes are used for strings.

2. Use those attributes where needed in our behaviors.

We need to set up the serial monitor to be used. We will do that in the setup function. Our syntax for it is above so we just follow it.

Serial.begin(<baud>);

Would be:

Serial.begin(9600);

```
void setup()
{
   // initialize digital pin 13 as an output.
   pinMode(pin13, OUTPUT);

   // Initialize Serial Monitor (more on this later)
   Serial.begin(9600);
}
```

Remember, we want to try to stay in line with the optimized code if possible. We can make use of the for loop that creates our blinks. Let's deal with theBlinks function first.

The variables dwDotPrintSym and dwDashPrintSym is going to be used in theBlink functions.

We need to add them where we all calling to the

```
/// S printing S Attribute(variable)
// .

theBlinks(dwDot, pin13, dwBetween, dwDotPrintSym);

/// O printing O Attribute(variable)
// -

theBlinks(dwDash, pin13, dwBetween, dwDashPrintSym);

/// S printing S Attribute(variable)
// .

theBlinks(dwDot, pin13, dwBetween, dwDotPrintSym);
```

function in the loop.

Remember since we are passing variables to the behavior function We have to add another variable to pass into it.

```
// Behaviors

int theBlinks(int passing1, int passing2, int passing3, char passing4)

{

        Instructions For Code Function

}
```

For the two remaining attributes(variables) dwDotPrintLet and dwDashPrintLet we need to think it out.

1. We don't need it to be repeated like our dots and dashes.

2. We want it to be printed at the end of the 3 dots and 3 dashes.

3. We would not place it in the "for" loop.

4. So for simplicity sake let's just add the code in the loop with the calls to theBlink function.

5. We would not put the code over the calls because we want the letters after the dots and dashes. That mean we will add it after each call as shown.

```
theBlinks(dwDot, pin13, dwBetween, dwDotPrintSym);
Serial.println(dwDotPrintLet);

/// O printing O Attribute(variable)
// -

theBlinks(dwDash, pin13, dwBetween, dwDashPrintSym);
Serial.println(dwDashPrintLet);

/// S printing S Attribute(variable)
// .

theBlinks(dwDot, pin13, dwBetween, dwDotPrintSym);
Serial.println(dwDotPrintLet);
```

We are using .println because we want it to start a new line after the letters are printed.

Back to our variables dwDotPrintSym and dwDashPrintSym.

3. Go into our behaviors and add the code needed to print to the serial monitor.

The behavior function we are dealing with is theBlink. We only need to add one line of code. We want the dots or dashes to be printed on the same line so we will use .print .

Serial.print(passing4);

Remember the loop is going to do the work of looping the code. That is why we only need to add one line. It will repeat 3 times according to the for loop.

We will add it after the code that was already there.

```
int theBlinks(int passing1, int passing2, int passing3, char passing4)
{

    for (int e = 0; e < 3; e++)
    {

        // turn the LED on (HIGH is the voltage level)
        digitalWrite(pin13, HIGH);
        delay(passing1);        // wait variables milliseconds
        // turn the LED on ( LOW is the voltage level)
        digitalWrite(pin13, LOW);
        delay(passing1);        // wait variables milliseconds

        // 4. Adding Serial Monitor funtions within loop
        Serial.print(passing4);
    }

}
```

4. Verify and upload

5. Troubleshoot as needed

6. Have a grape(much healthier than a cookie).

So did you work through the algorithm like above, or did you find another way to do it?

Checking The Sketch

A part of troubleshooting this sketch was to open the serial monitor and check what printed to the screen. That is how you would check to make sure it is printing correctly.

What should show on the serial monitor should be:

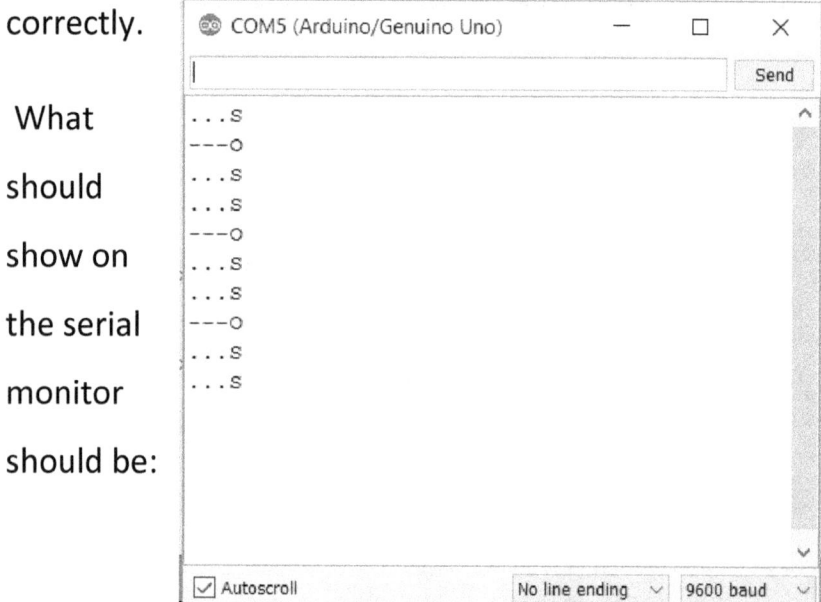

The SOS will repeat over and over. Just like the blinks.

The code to use for the next chapter is below.

```
// morseSos.ino

// Attributes

const int pin13 = 13;      // Stores pin number

const int dwDot = 300;    // Stores 300 millisec value

const int dwDash = 900;   // Stores 900 millisec value

const int dwBetween = 1000;   // Stores 1 seco value



/* Attributes to print to serial monitor

 *  For dot and dash. Then S and O.

 */



const char dwDotPrintSym = '.';
```

```
const char dwDashPrintSym = '-';

const char dwDotPrintLet = 'S', dwDashPrintLet = 'O';



/*

 setup function runs once when you press reset or
power the board

*/

void setup()

{

    // initialize digital pin 13 as an output.

    pinMode(pin13, OUTPUT);


    // Initialize Serial Monitor (more on this later)

    Serial.begin(9600);

}
```

```
// the loop function runs over and over again forever

void loop()

{

    // .

    /// S printing S Attribute(variable)


    theBlinks(dwDot, pin13, dwBetween,
dwDotPrintSym);

    Serial.println(dwDotPrintLet);



    // -

    /// O printing O Attribute(variable)



    theBlinks(dwDash, pin13, dwBetween,
```

```
dwDashPrintSym);

  Serial.println(dwDashPrintLet);



  // .

  /// S printing S Attribute(variable)

  theBlinks(dwDot, pin13, dwBetween,
dwDotPrintSym);

  Serial.println(dwDotPrintLet);

}


// Behaviors

int theBlinks(int passing1, int passing2, int passing3,
char passing4)

{



      for (int e = 0; e < 3; e++)
```

```
{

    digitalWrite(pin13, HIGH);

    delay(passing1);

    digitalWrite(pin13, LOW);

    delay(passing1);

    /*

    4. Adding Serial Monitor funtions within loop

    */

    Serial.print(passing4);

}

// delay 1 seconds between letters

delay(passing3);        // wait for a second

}
```

Summary:

This chapter is really a bit of a recap from project 1.

Serial Monitor is the only new thing added. You will find that you will use it a lot for many different things.

If you didn't quite get it, study the code above, or get the book mentioned above where this code is built out. If you got it then let's continue on.

But what about optimizing the code and making the code simple to read.

The blinks is vey easily read and optimized. The code we added for the serial monitor is valid. It is not so easily read or optimized.

If statements is going to be introduced in the next chapter. It should solve our problem of optimization and readibility.

Have a grape and meet me there.

2 CHAPTER IF STATEMENT

The if statements allows your program to make decisions. It test the conditions you set. It do or don't do what you have for it's instructions.

The syntax is:.

    if(<condition>)

    {

        <Instructions for the function>

    }

The condiition in the parameters is how you test code. That is how your program can make decisions.

Inside the parameters(condition) is true if you have only one variable. If you add a variable in the parameters in the if statement it will evaluate to true.

Using the syntax above on the sketch, let's do an if statement.

Algorithm:

1. If led delay for 300 milliseconds.

2. Print . . .

3. S same line but after it start a new line.

That is the algorithm.

Remember our blinks are fine so we don't have to do anything with them. It is our printing to the serial monitor we are working on.

Parameters of a function can be very good for a trigger for the if statement. Look at your function carefully.

What do you mean by trigger? You ask.

I mean what is the best variable we can use to trigger the if statement. What variable passed in our params is allowing the led to blink for the dot and dashes? It must be something that is true about your code.

That would be dwDot and dwDash which is passed to passing1 for our blinks. It is true that for dots it is 300 milliseconds defined in the variable definition.

We now have or condition.

    If(passing1)

Now we give our if statement instructions in it's curly brackets, just like any other function.

That is 2 of the algorithm.

Then let's do 3 of the algorithm too.

```
if(passing1)

{

    Serial.print(" . . . ");    // prints . . .

    Serial.println("S");

    // prints S then start a new line.

}
```

The if statement goes in the same function with your blinks.

When you verify and upload your code you should get something like below. Check your serial monitor. Check your led blinks are working.

```
delay(dwBetween);               // wait for a second

if(passing1)
{
   Serial.print(" . . . ");   // prints . . .
   Serial.println("S");       // prints S then start a new line.
}
```

Don't worry there is no error. There is still the original code for the serial monitor done in chapter 1.

Can you do the same for O now?

If you feel you can then do the same thing for O. Delete the code for serial print that is not in the if statements.

Did you say No?

If you feel you can't continue on.

Check Your First If Statement

Did you say yes or no?

For those that said yes and found a way. More power to you. Good job.

My answer would have been no. Good job for those that agree.

Why No? You ask.

It would only show S because we are passing variables to the function. S is the first pass. O would not show.

We don't have to do without O though. The if statement can handle this.

How did you do?

Check your code against the code below.

Remember that your blinks should be flashing correctly too.

```
// morseSos.ino


// Attributes

const int pin13 = 13;      // Stores pin number
```

```
const int dwDot = 300;    // Stores 300 millisec value

const int dwDash = 900;   // Stores 900 millisec value

const int dwBetween = 1000;   // Stores 1 sec value



/*

 setup function runs once when you press reset or power
the board

*/

void setup()

{

    // initialize digital pin 13 as an output.

    pinMode(pin13, OUTPUT);


    // Initialize Serial Monitor (more on this later)

    Serial.begin(9600);

}
```

```
// the loop function runs over and over again forever

void loop()

{

    // .

    /// S printing S Attribute(variable)


    theBlinks(dwDot, pin13, dwBetween);


    // -

    /// O printing O Attribute(variable)


    theBlinks(dwDash, pin13, dwBetween);


    // .
```

```
    /// S printing S Attribute(variable)


    theBlinks(dwDot, pin13, dwBetween);

}


// Behaviors


int theBlinks(int passing1, int passing2, int passing3)

{

    for (int e = 0; e < 3; e++)

    {

        digitalWrite(pin13, HIGH);

        delay(passing1);      // wait variables
milliseconds

        digitalWrite(pin13, LOW);

        delay(passing1);      // wait variables
```

milliseconds

```
    // 4. Adding Serial Monitor funtions within loop

}


    // delay 1 seconds between letters

    delay(passing3);          // wait for a second


    if(passing1)

    {

        Serial.print(" . . . ");  // prints . . .

        Serial.println("S");

        // prints S then start a new line.

    }

}
```

The results of the code for serial monitor above should

be:

```
delay(dwBetween);                    // wait for a second

if(passing1)
{
  Serial.print(" . . . ");  // prints . . .
  Serial.println("S");   // prints S then start a new line.
}
```

Logical Operators

There are operators that you use with variables to test the conditions. Which takes it further.

You already know operators. You have been using them since your first day in school when you started math.

Addition   +

Subtraction   -

Multipication   *

Division   /

Equal   =

Tip: Equal is also an assignment operator as when defining variables.

For example:

    Int variable1 = 3;

For the if statements there is Logical Operators. Logical Operators is what you use with the if statement.

For basic if testing.

    If (<variable1> <logical operator> <logical operator><varable2 or number>)

The if statement will compare variable1 and variable2 or a number.

The true side is the left side, It should be used for variables.

I think of it as the left hand side or the right hand side.

Like in: give me five   on the left had side.

That is my ebony speaking.

Sometimes you have to figure out things in your mind the way you will understand it for programming.

With the if statement you want your programs to make decisions. That is what make a program intelligent. If they can make the right decisions you can depend on them to do it right every time.

Unfortunately, the if statement can be used wrong(and very easily let me tell you). In that case if will indeed make your program do some dumb things.

Don't let that make you shy away from if statements. Correctly used it can help in optimization and reading code.

If (<variable1> <logical operator> <logical operator><varable2 or number>)

Looking at the above line variable1 is the leftHand side(the true side). Variable2 or number is the rightHand side(the test side). Your operator or operators go in the middle.

Keep in mind you can use numbers or variables, but most often it should be variables. Numbers are fine for the rightHand side.

Left and Right plays a big part in this.

Let's go over the simple operators you can use to work with the two sides. We are going to add examples using the variables for delay. In our code it is passing1.It will either be 300 or 900.

==   True

This means if the rightHand side matches with the lefthand(true side) side the condition is true and the code will run.

Knowing that dwDot is 300. The right hand side has to match with 300 or it will be the condition test false.

We already have the if statement done for S. Using == we can test the passing 1 variable.dwDot is passed to theBlink function.

We change it like so:

```
if(passing1 == 300)

    {

        Serial.print(" . . . ");  // prints . . .

        Serial.println("S");

        // prints S then start a new line.

    }
```

This gives the condition a test to do. If the right hand side shows anything by 300, the condition will be false and the code in the curly brackets would not run.

Once you have made the change to the condition upload your code and very it. You Should get the same results as before.

Listing the other Operators.

!=   not equal, false, or opposite.

If you change the == in your if statement to != , the

instruction code will not run for 300. It means that anything but 300 will run your instruction code.

If you check your serial monitor you are going to see the code run, but it is running for the O not the S.

More operators are:

> greater than   Greater than the leftHand side.

< Less than   Less than the LeftHand side.

>= greater than or equal to   The leftHand side

<= Less than or equal to   The leftHand side

&&   and   both leftHand side and righthand side are true.

|| or   either left hand side or righthand side is true

!   not   false

To the beginners eye all of this can be confusing.

Just remember that the left hand side is true. The right hand side is the test side.

Algorithms will help you sort out what you need to do.

The if statement is something that you need to test a lot to figure out how to use the above logical operators.

When your are experimenting make sure you know why the if statement is doing what it is doing.

If else

We have our code but we only have one set of instructions to use for the code which can be very confusing.

If else allows you to test a condition. Add a set of instructions if the condition is true, else another set of instructions if the condition is false.

This should help with the code that we are running.

The syntax is:

```
if(<condition>

{

        // If condition is true

        <Instructions for the function>
```

```
    }

    else

    {

        // If condition is false

        <Instructions for the function>

    }
```

Going back to our code as we left it with != as the logical operator. The else will be added to the if statement.

Example 1:

if(passing1 != 300)
 	
{
 	
    Serial.print(" . . . "); // prints . . .
 	
    Serial.println("S"); // prints S then start a new line.
 	
}
 	
else
 	
{

```
Serial.print(" - - - ");  // prints . . .

Serial.println("O");  // prints S then start a new line.

}
```

Although the if else statement is set up correctly it is still not runing correctly.

```
if(passing1 != 300)
{
    Serial.print(" != 300. Should be O");
    Serial.print(" . . . ");  // prints . .
    Serial.println("S");  // prints S then
}
else
{
    Serial.print(" == 300. Should be S");
    Serial.print(" - - - ");  // prints . .
    Serial.println("O");  // prints S then
}
```

Why?

Remember: if else can give you a lot of advantages if done correctly. This is where attention to detail really counts.

```
COM5 (Arduino/Genuino Uno)
|
== 300. Should be S - - - O
!= 300. Should be O . . . S
== 300. Should be S - - - O
== 300. Should be S - - - O
!= 300. Should be O . . . S
== 300. Should be S - - - O
== 300. Should be S - - - O
!= 300. Should be O . . . S
== 300. Should be S - - - O
== 300. Should be S - - - O
!= 300. Should be O . . . S
== 300. Should be S - - - O
```

It will run correctly if we change the != (not equal) to ==(equal to). Because that will mean that dwDot is equal to 300.

Example 2:

```
if(passing1 == 300)

{

    Serial.print(" . . . ");  // prints . . .

    Serial.println("S");  // prints S then start a new line.

}

else

{

    Serial.print(" - - - ");  // prints . . .

    Serial.println("O");  // prints S then start a new line.

}
```

Or

We can leave it with the !=(not equal) and change the instructions in the brackets for the If else statement to;

```
SosifStatement a
if(passing1 == 300)
{
    Serial.print(" == 300. Should be S");
    Serial.print(" . . . ");  // prints . .
    Serial.println("S");  // prints S then
}
else
{
    Serial.print(" != 300. Should be O");
    Serial.print(" - - - ");  // prints . .
    Serial.println("O");  // prints S then
}
```

```
COM5 (Arduino/Genuino Uno)

== 300. Should be S . . . S
!= 300. Should be O - - - O
== 300. Should be S . . . S
== 300. Should be S . . . S
!= 300. Should be O - - - O
== 300. Should be S . . . S
== 300. Should be S . . . S
!= 300. Should be O - - - O
== 300. Should be S . . . S
== 300. Should be S . . . S
!= 300. Should be O - - - O
== 300. Should be S . . . S
```

Example 3:

```
if(passing1 != 300)

{

    Serial.print(" - - - ");  // prints . . .

    Serial.println("O");  // prints S then start a new line.

}

else

{

    Serial.print(" . . . ");  // prints . . .

    Serial.println("S");  // prints S then start a new line.

}
```

```
if(passing1 != 300)
{
  Serial.print(" != 300. Should be O");
  Serial.print(" - - - ");  // prints . .
  Serial.println("O");  // prints S then
}
else
{
  Serial.print(" == 300. Should be S");
  Serial.print(" . . . ");  // prints . .
  Serial.println("S");  // prints S then
}
```

That will mean that O will run for anything but 300.

COM5 (Arduino/Genuino Uno)

```
== 300. Should be S . . . S
!= 300. Should be O - - - O
== 300. Should be S . . . S
== 300. Should be S . . . S
!= 300. Should be O - - - O
== 300. Should be S . . . S
== 300. Should be S . . . S
!= 300. Should be O - - - O
== 300. Should be S . . . S
== 300. Should be S . . . S
!= 300. Should be O - - - O
== 300. Should be S . . . S
```

Either Example 2 or Example 3 will work. But they are incomplete. The condition set for dwDot is ok. But it leave dwDash open to be true for anything except for 300. It should only be true to 900.

You want your code to be TIGHT. With no room for errors.

It is just a matter of taking it a little further.

Adding else if To The Mix

It is really - if else if. - That is how we will TIGHTEN up the if statement. That is important so that if something other then 300 is added to the code later O will only run for it's delay time of 900. We are avoiding bugs this way.

It is really simple to do since we have come this far. We are going to use example 2 from above.

Adding else if.
if(passing1 == 300)

```
{

    Serial.print(" . . . ");  // prints . . .

    Serial.println("S");  // prints S then start a new line.

}

else if(passing1 == 900)

{

    Serial.print(" - - - ");  // prints . . .

    Serial.println("O");  // prints S then start a new line.

}
```

Now the code is set to run correctly for the blinks in the serial monitor.

```
        delay(dwBetween);                        // wait fo

if(passing1 == 300)
{
    Serial.print(" == 300. Should be S");
    Serial.print(" . . . ");   // prints . .
    Serial.println("S");   // prints S then
}
else if(passing1 == 900)
{
    Serial.print(" != 300. Should be O");
    Serial.print(" - - - ");   // prints . .
    Serial.println("O");   // prints S then
}
```

You can further tighten it up by adding another else to

this if statement. This one to print an error message. It will run to that message if the value is not 300 or 900.

```
 COM5 (Arduino/Genuino Uno)

        300. Should be S . . . S
!= 300. Should be O - - - O
== 300. Should be S . . . S
== 300. Should be S . . . S
!= 300. Should be O - - - O
== 300. Should be S . . . S
== 300. Should be S . . . S
!= 300. Should be O - - - O
== 300. Should be S . . . S
== 300. Should be S . . . S
!= 300. Should be O - - - O
```

Checking The Sketch

The code is below:

// morseSos.ino

// Attributes

const int pin13 = 13;   // Stores pin number

```
const int dwDot = 300;    // Stores 300 millisec value

const int dwDash = 900;   // Stores 900 millisec value

const int dwBetween = 1000;   // Stores 1 sec value



/*

setup function runs once when you press reset or power
the board

*/

void setup()

{

    // initialize digital pin 13 as an output.

    pinMode(pin13, OUTPUT);



    // Initialize Serial Monitor (more on this later)

    Serial.begin(9600);

}
```

```
// the loop function runs over and over again forever

void loop()

{

    // .

    /// S printing S Attribute(variable)


    theBlinks(dwDot, pin13, dwBetween);



    // -

    /// O printing O Attribute(variable)


    theBlinks(dwDash, pin13, dwBetween);



    // .
```

```
    /// S printing S Attribute(variable)


    theBlinks(dwDot, pin13, dwBetween);

}


// Behaviors


int theBlinks(int passing1, int passing2, int passing3)

{

    for (int e = 0; e < 3; e++)

    {

        digitalWrite(pin13, HIGH);

        delay(passing1);

        digitalWrite(pin13, LOW);

        delay(passing1);
```

```
}


    // delay 1 seconds between letters

    delay(passing3);          // wait for a second

// 4. Adding Serial Monitor funtions within loop

    if(passing1 == 300)

    {

        Serial.print(" == 300. Should be S");

        Serial.print(" . . . ");  // prints . . .

        Serial.println("S");

// prints S then start a new line.

    }

    else if(passing1 == 900)

    {

        Serial.print(" == 900. Should be O");

        Serial.print(" - - - ");  // prints . . .
```

```
        Serial.println("O");

    // prints S then start a new line.



    }

else

{

Serial.println(" Error: Value not found ");

}



}
```

Summary:

Verify and upload. The code is now complete and optimized for the printing of the dots, dashes, S, and O.

It is all in the if statement and very readable. It only runs when the conditions are true.

That is what you want. As your program grows you want to keep it as compact as possible. At the same time you want it to do what you need it to do.

We could have written a whole print function. But for this program it is not needed.

Reusing code is key and we reused the passing 1 code in the if statement.

3 CHAPTER TROUBLESHOOTING CODE

No matter what you do there will be times when a block of code is just not doing what you want. The IDE will verify and it will upload but it is not doing what you want. No amount of hair pulling, banging on desk, straining your eyes to make the problem show will help.

Sometimes you just need to take a break from the code. Take a nap. Go for a ride. Eat a cookie.

But, keep pushing on. Don't give up on it. Some of the hardest errors leads to ah-ha moments. Those ah-ha moments are very rewarding.

But, remember now you have the serial monitor. It allows you to communicate with your Uno.

Serial Monitor To Troubleshoot

Maybe you didn't notice but we used the serial monitor to troubleshoot in the last chapter. Below is an example to show you.

```
⑆eslfStalemonIfa

if(passing1 != 300)
{
    Serial.print(" != 300. Should be O");
    Serial.print(" . . . ");  // prints .
    Serial.println("S");  // prints S then
}
else
{
    Serial.print(" == 300. Should be S");
    Serial.print(" - - - ");  // prints .
    Serial.println("O");  // prints S then
}
```

The block of code was:
if(passing1 != 300)
{

```
    Serial.print(" . . . ");  // prints . . .

    Serial.println("S");  // prints S then start a new line.

}

else

{

    Serial.print(" - - - ");  // prints . . .

    Serial.println("O");  // prints S then start a new line.

}
```

As you can see in the picture there was extra code added within the brackets of the if statement to show what was supposed to be printed.

```
    Serial.print(" == 300. Should be S");
```

and

```
    Serial.print(" == 900. Should be O");
```

Both of those lines were added for the purpose of troubleshooting.

When it didn't match up we knew where the problem was. Problem Solved.

You will know what you want your code to do and what point you want your code to do it.

Add a print or a println where you want a block of code to run. A short line saying what you want it to do is all that is needed.

As you see from the picture the code was not running the way it should. It was easily corrected.

This can be done in any functions you need it to be done in. Once your code is running correctly delete it.

COM5 (Arduino/Genuino Uno)

```
==  300.  Should be S - - - O
!=  300.  Should be O . . . S
==  300.  Should be S - - - O
==  300.  Should be S - - - O
!=  300.  Should be O . . . S
==  300.  Should be S - - - O
==  300.  Should be S - - - O
!=  300.  Should be O . . . S
==  300.  Should be S - - - O
==  300.  Should be S - - - O
!=  300.  Should be O . . . S
==  300.  Should be S - - - O
```

Using the serial monitor to check functions, variables, even arrays(we will get to that later) can be a time saver.

If statement to troubleshoot

If statements used correctly can also be used to

troubleshoot.

One way is to check to make sure your variables are correct. Again checking elements in arrays to see if it is correct. It is good for testing conditions.

Using if statements to troubleshoot is just... logical. It can also give you extra practice for when you need it for your program.

Google, YouTube, And Forums

This probably goes without saying but remember it well. There is a whole world and community of teaching and assistance. When it come to programming the help is out there. Especially if you are self taught.

You may not be able to find the exact answers you are looking for but help is out there. When trying to find your answers have patience and be open minded . Sometimes the answer is right there and your just not seeing it right.

You can copy errors showing in the Arduino IDE screen. Paste them right into your search engine. Sometimes you luck up on a solution right away. More often you have to dig for it.

Books sometimes is good. If you are a visual learner YouTube is the place for you. Some people put out 100's of videos (and good ones) to instruct.

Ebonygeek45 is trying to get to a point to put out good videos in a timely manner. As it grows this will happen.

When dealing with electric components you may want to look up data sheets. Google pictures are sometimes good for that. Learning how to read data-sheets for electric components is another aspect to Arduino. But it is easier then the programming.

Forums are very good. If you just can't work your way through an error post it on a forum. Make sure the forum is for C++ and better yet for Arduino. Those guys are good at showing you where you went wrong. Some of them are crabby. Some point you to a forum where your problems has already been solved. Even the crabby mean ones will sometimes stick with your post until your problem is solved.

They will give you opinions. Remember It is easier to catch bees with honey then vinegar. Take what you can use and let the rest goes. They are giving of their time to help. They do help.

Sometimes if they get the idea that you are a student they will be less likely to help. That is because they don't want those students to have them doing there homework. They know it is not doing them any favors to give them the answers.

Document Your errors

Put together some cheat sheets for your syntax.

Keep a journal or document with the errors you run into and how you solved them. When you run into the problem again. It may be as simple as pulling up that document and using find to find the error. You would have solved your own problem.

Keeping your own document of error is probably one of the best things you can do. It is a way to help yourself.

Summary:

The fact is that you will be doing a lot of troubleshooting. That is just the nature of the beast. Even seasoned programmers do it, they just know how to get to there answers better.

If you keep with it, you will find that you will spend less time
on it. You will make less mistakes.

4 CHAPTER SWITCH AND CASE

Switch and Case is very like the if statement . What you

can do with an if statement you can do with a Switch and Case mostly.

It boils down to preference really.

If you find that an if statment is getting long, you may want to do a switch instead.

Why? You ask.

Because whether then testing a lot of different conditions with the if statement, you can test multiple conditions with the switch. The advantages start to show when you are testing more then 3.

They are similiar because they both test conditions.

The syntax is different and the readability is better.

Just like with the if statement you want to find your trigger.

Once you have your trigger

You add it to the parameters of the switch.

Keep in mind the type for the switch should be int, long, or char.

Using The Switch And Case

The syntax for the switch is:

The syntax is:

```
switch(<condition>)

{
        case <value>:  <code to run>

                       <code to run>

                       break;

        case <value>:  <code to run>

                       <code to run>

                       break;

        default: Serial.println("<default messege if case
doesn't match>");

                       break;

}
```

Using our example we did for the if Statement.

You can either delete the if stament or comment it out.

We are using passing1 for the blinks. Just like with the if statement we are going to add passing1 for the conditions in the parameters for the switch.

When the code runs to the switch. It is going to see tha passing1 variable. It stores the 300 for dots or the 900 for dashes.

The passing1 variable is a placeholder for dwDot and dwDash. Again passing1 is the trigger.

The first case will be 300 and the second case will be 900. Depending on which variable passing1 is holding a place for, it will compare that value for the variable to the case value.

Do not leave a space between the case value number and the colon. If you do you will get symbols that have nothing to do with your code.

Next just add the code to run for 300 then for 900.

The swich and case should look like below:
switch(passing1)

```
{
        case 300:    Serial.print(" . . . ");

                 Serial.println("S");

                 break;

        case 900:    Serial.print(" - - - ");

                 Serial.println("O");

                 break;

        default: Serial.println("Error: Value Not
Found");

                 break;

}
```

The code for the switch statement is placed where the if statement was.

Upload it then verify it. Pull up your serial monitor and check that it ran correctly. Check you led to make sure it is blinking correctly.

Explaining The Switch Further

If the condition variable matches with the case value, it will run the code for that case.

The break added after the code to run stops the switch and exit out of it.

The default is for if the condition value does not come to a case it matches. It runs the default code. When it get to break it exit out of the switch.

If you don't add break at the end of a case. the switch will continue and run the code for the next case until it get to break. Sometimes that can be helpful. Here we are using the break at the end of every case.

Checking The Sketch

It wasn't so much different from the if statement. It did the same thing.

Compare the if statemnt with the switch to see the difference between them.

```
// morseSos.ino
```

```
// Attributes

const int pin13 = 13;       // Stores pin number

const int dwDot = 300;    // Stores 300 millisec value

const int dwDash = 900;   // Stores 900 millisec value

const int dwBetween = 1000;   // Stores 1 sec value



/*

 setup function runs once when you press reset or power
the board

*/

void setup()

{

    // initialize digital pin 13 as an output.

    pinMode(pin13, OUTPUT);



    // Initialize Serial Monitor (more on this later)
```

```
    Serial.begin(9600);

}



// the loop function runs over and over again forever

void loop()

{

    // .

    /// S printing S Attribute(variable)

    theBlinks(dwDot, pin13, dwBetween);

    // -

    /// O printing O Attribute(variable)

    theBlinks(dwDash, pin13, dwBetween);

    // .

    /// S printing S Attribute(variable)

    theBlinks(dwDot, pin13, dwBetween);
```

```
}


// Behaviors

int theBlinks(int passing1, int passing2, int passing3)

{


    for (int e = 0; e < 3; e++)

    {

        digitalWrite(pin13, HIGH);

        delay(passing1);

        digitalWrite(pin13, LOW);

        delay(passing1);

    }
```

```
// delay 1 seconds between letters

delay(passing3);          // wait for a second

// 4. Adding Serial Monitor funtions within loop

switch(passing1)

{

    case 300:   Serial.print(" . . . ");

                Serial.println("S");

                break;

    case 900:   Serial.print(" - - - ");

                Serial.println("O");

                break;

    default:    Serial.println("Error: Value Not
Found");

                break;

    }

}
```

Your serial monitor should show the below results.

Optimizing Code

We are going to stick with the Switch. To take it a little further lets put the Switch in another function.

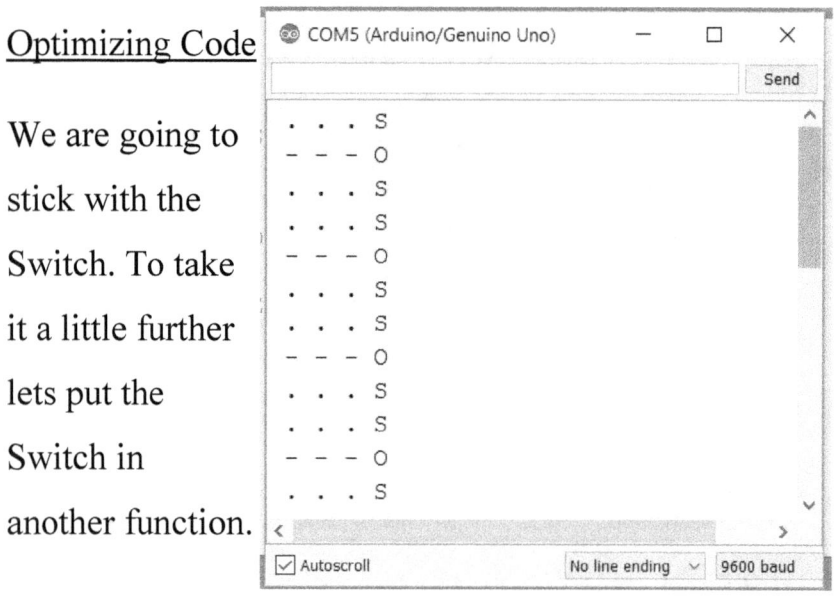

This will be a big improve from how the code look at the end of chapter 1.

We went over passing variables to functions in the first

book. This should be something that we know already.

In this case we are going to use passing1 as the variable we pass to this function. We will name it printSos() .

This means that passin1 is going to be reused again.

You can never go wrong making your code resuable.

Remember if we ever change the delay time for the variable we just change it where the variable is defined.

If you want to take it even further you can add the case values in a variable too. For now we are going to leave them as is.

Checking The Sketch

Your code should now be.

// morseSos.ino

// Attributes
 	
const int pin13 = 13;   // Stores pin number
 	
const int dwDot = 300;   // Stores 300 millisec value
 	
const int dwDash = 900; // Stores 900 millisec value

```
const int dwBetween = 1000;   // Stores 1 sec value



/*

setup function runs once when you press reset or power
the board

*/

void setup()

{

    // initialize digital pin 13 as an output.

    pinMode(pin13, OUTPUT);


    // Initialize Serial Monitor (more on this later)

    Serial.begin(9600);

}


// the loop function runs over and over again forever
```

```
void loop()

{

    // .

    /// S printing S Attribute(variable)

    theBlinks(dwDot, pin13, dwBetween);

    // -

    /// O printing O Attribute(variable)

    theBlinks(dwDash, pin13, dwBetween);

    // .

    /// S printing S Attribute(variable)

    theBlinks(dwDot, pin13, dwBetween);

}


// Behaviors

int theBlinks(int passing1, int passing2, int passing3)
```

```
{

    for (int e = 0; e < 3; e++)

    {

        digitalWrite(pin13, HIGH);

        delay(passing1);

        digitalWrite(pin13, LOW);

        delay(passing1);

    }


    // delay 1 seconds between letters

    delay(passing3);          // wait for a second

    printSos(passing1);

}


int printSos(int passing1)
```

```
{



    switch(passing1)

    {

        case 300:   Serial.print(" . . . ");

                    Serial.println("S");

                    break;

        case 900:   Serial.print(" - - - ");

                    Serial.println("O");

                    break;

        default:    Serial.println("Error: Value Not
Found");

            break;

    }

}
```

```
    return passing1;



}
```

Summary:

Are you beginning to see the pattern:

1. Write up a Algorithm to show what need to be done

2. Work through the Algorithm until the code is working.

3. Examine the code

4. Check to see if repeating code can be improved.

5. Check to see if the code is not easily read.

6. If so start the process over again at 1 for the optimization.

7. When you are happy with it continue on with building it up more if needed.

8. If so start back over again at 1.

The pattern goes round and round don't it. In the end you will have a compact program with less repeating code(let's face it some of it will have to repeat), easily read with reusable code. In other words neat, readable, optimized code for your program.

5 CHAPTER LOOPS: WHILE AND DO WHILE

The loop we have been using in the code is the "for" loop. That has already been explained.

There is still the while and do while loop.

Which loop you decide to use in your programs is up to your preference.

In the case of Arrays the for loop is recommended.

Alll the loops pretty much do the same thing.

All will need a:

1. Start count set with a variable.

2. Condition inside their parameter.

3. Increment to keep tally of how many times looped.

The while Loop

The while and for loop are most similiar. The difference is how it is set up.

The while loop shown here is for our code. It should look most familiar to your eyes.

```
1   int e = 0;

2   while(e < 3)
    {
3       // turn the LED on (HIGH is the voltage level)
        digitalWrite(pin13, HIGH);
        delay(passing1);        // wait variables milliseconds
        // turn the LED on ( LOW is the voltage level)
        digitalWrite(pin13, LOW);
        delay(passing1);        // wait variables milliseconds

4       e++;
    }
```

The syntax is :

```
<type> <varName> = <startNumber>;

while(<condition>)

{

    <Instructions for code>;

    <variable increment>;

}
```

The Algorithm for the while loop was

- Repeat block of code for flashes 3 times.

  - Use while loop

    - 1. Declare and Define variable to increment.

    - 2. @ while params(parameters) test variable for true or   false.

- 3. Block of code to be executed if condition is true in   params. (Inside curly brackets)

- Variable increment. (Inside curly brackets)

Do you understand what that means?

Let's clarify a little.

The variable is defined as

the variable that will increment from 1 to 3(for the code done here).

Entering the while loop the conditions test the variable for true.

If it is set to start at 0 it will be true until it reach 3. Once it reach 3 it will test false and the loop will end.

In the curly brackets is the code to run as long as the condition is true.

The variable to increment is placed after the code to run, and it will increment by 1. Again when it increment three times the loop will end.

It should now be simple to see how similiar to the for loop the while loop is.

It is placed where the for loop was placed in the sketch.

The do while Loop

The do while loop is the least used, and I would say most misunderstood loop, although it should not be.

The do while loop test the condition of the loop last. That is what set it apart from the "for" and "while" loop.

The do while loop is shown below. Even though it looks a bit different it requires the same information as the othe two loops.

```
1  int e = 0;

   do
   {
2      e++;
       // turn the LED on (HIGH is the voltage level)
       digitalWrite(pin13, HIGH);
       delay(passing1);        // wait variables milliseconds
3      // turn the LED on ( LOW is the voltage level)
       digitalWrite(pin13, LOW);
       delay(passing1);        // wait variables milliseconds
   }
4  while(e < 3);
```

The syntax is :

```
<type> <varName> = <startNumber>;

do

{

    <variable increment>;

    <Instructions for code>;

}

while(<condition>)
```

The Algorithm for the while loop was

- Repeat block of code for flashes 3 times. (For some reason)The first block of could should run one time no matter what.

  - Use do while loop

    - 1. Declare and Define variable to increment.

    - 2. Variable increment. (Inside curly

brackets)

- 3. Block of code to be executed the first time without test. The remaining times the block of code will test if condition is true in params. (Inside curly   brackets).

- 4. @ while params(parameters) test variable for true or   false at the end of the loop..

To clarify the do while loop.

The variable is defined for the variable that will increment from 1 to 3(for the code done here).

Entering the do part of the loop There are no parameters to test the condition

Inside the curly brackets the variable increments first.

Then the instructions for the code to be ran in the loop.

It comes to the closing curly bracket

The While loop is under it with the condition to be tested.

It is tested at the end of the loop. If it test true it will loop back to the do part . Iw will continue on that way until the condition shows false.

Ebonygeek45 speak: The do wile loop is like a while loop upside down,

You can think of it that way if it helps.

Summary:

The for loop was already explained in depth in the previous book.

It is the one that we are using in our code. So go ahead and replace the do while loop with the for loop we been using all this time.

That is all for loops.

Simply put, if you want the block of code to run once no matter what. Then test the condion at the end of the loop. You would use a do while loop.

If you want the Block of code to be tested by it's

conditions at the beginning of the loop, and if true run the code. You would use the <u>while</u> or <u>for</u> loop.

Again for an array or container of some kind, use the <u>for</u> loop.

Most prefer to use the for loop because required code for it is all in the parameters. There is less chance of forgetting because it is all in one place.

There is a place for the do while loop. The do while will always loop once even if the condition is never met.

Sometimes you may need the code to run a least once before testing the conditions(very rarely). In that case you would use <u>do while.</u>

6 CHAPTER TESTING CODE WITH NUMBERS

You may have noticed that we have not been changing the code much in this book.

We have been going through ways you can work with your code. Ways you can make your code smart. Still this is just scratching the surface. Because there are many ways to work with your code. You will develop your own style. It is good to know all your options so you can develop a productive style.

Understanding how you can work with your code makes it easier if you use someone elses sketch or library. It also shows what is possible when building your own programs. Still this is basic and a bit of intermediate skill sets.

You have just been taken through:

1. if, else if, else   statements

2. switch and case

3. while, do while   loops

Those are ways to make your code do something. Whether that be to:

1. Make a decision based on input or output.

2. Go through a block of code multiple times.

These are ways to make the code smart. The way a programs should be. Remember that it is based on your code so the right syntax is crucial. Your program can behave a bit dumb and confusing if you do not program it right.

We have also worked with the serial monitor. That is another way to have the uno communicate back to you based on your coding.

There will be times when you are not sure if what you are doing is working the way you would like. There are many ways to try to figure it out. Try – Try – and Try again is a way.

But, have you ever tried looking it up on the internet and see code for circles and squares, or numbers?

You may think that will not help you. But, sometimes it does help to try it the way the article on the internet, a book, video, or even the Arduino example sketch is showing is showing.

Pass By value Test

For instance passing variable values in params is straightforward. Especially if the code work right.

Did you realize that you can test code in the Arduino IDE as long as your uno is plugged in. This can be code that have nothing to do with the uno. That can allow you to test something to understand it before you add it to your program.

That is what is going to be shown now. To get the most out of this example follow along.

You should already know that whenever you want to start a new sketch you click:

File > New

This will bring up a new sketch.

Will save it to whatever name you choose:

File > Save As...

For this example I will name it dummy1PassValue. It is not a sketch that will have uno programming in it. Thus it will be a dummy file. It is named so I will know it is a dummy file with a pass by value test.

My main code is going in the loop. My functions are going underneath the loop.

Passing by value using numbers.

Add two interger attributes:

    5 and 10

This is a dummy test so we are adding them in the loop.

```
1  void setup() {
2      // put your setup code he
3
4  }
5
6  void loop()
7  {
8      // put your main code her
9
10     // Attributes
11     int a = 5;
12     int b = 10;
13
14  }
```

Adding it at the top is fine too.

Remember up until now we have wanted the params(parameters) to act as a placeholder for when we call the function. Then when we call the function we add the variable we wanted to use. The value of the variable is then placed where the place holder variable was.

It does exactly what we want it to do.

The serial monitor is being used to make the text printed to the serial monitor more easily read. Also, to explain what is being done.

For example the sketch below:
/*   dummy1PassRef.ino   */

void setup()
 	
{
 	
    // put your setup code here, to run once:

```
    Serial.begin(9600);     // To view serial monitor

}

void loop()

{

    // put your main code here, to run repeatedly:

    // Attributes

    int a = 5;

    int b = 10;

    passByValue(a);

    Serial.println("_____");

    passByValue(b);

    Serial.println("_____");

}

int passByValue(int placeholder)

{
```

```
// Using if statement to test the attributes.

if(placeholder == 5)

{

    Serial.println( "This should be 5 for A: " );

    Serial.println(placeholder);

}

else if(placeholder == 10)

{

    Serial.println( "This should be 10 for B: " );

    Serial.println(placeholder);

}

}
```

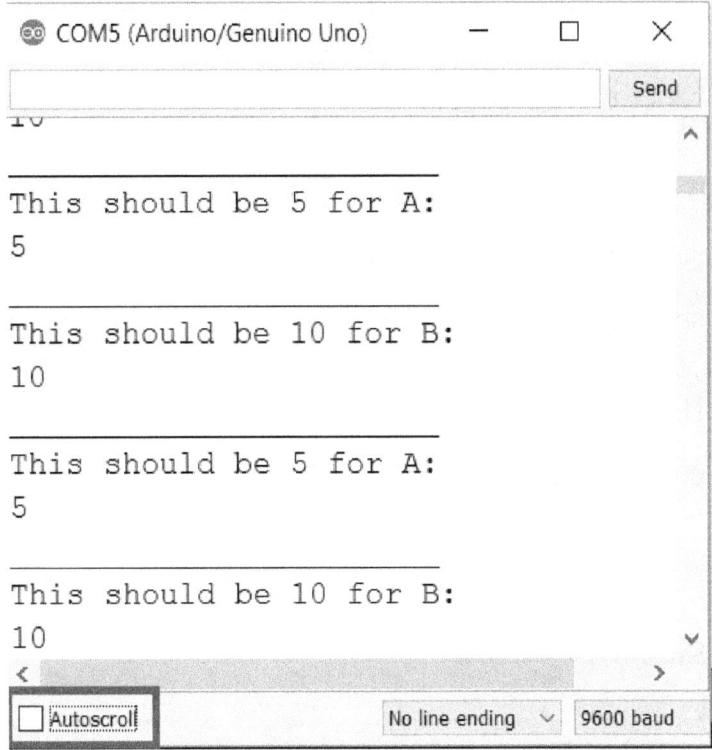

Try It

Your serial monitor should show that it is passing the value correctly.

If your serial monitor is running to fast, uncheck Autoscroll at the bottom.

This shows that Passing by value copies the value to the function it is passed to.

You should understand the sketch because it is using the serial monitor and if statement as previous chapters have shown.

If the sketch confuse you go back to chapter 1 and chapter 2 and refresh.

Summary:

Using numbers to see how code behaves with them first can be useful.

Once you have in your mind how the code is supposed to behave, you can add it to your project with confidence.

The sketch above shows how passing by value works in a simplified way.

If you now take a look at the sos code you should see how it is the passing by value for theBlinks.

7 CHAPTER PASS BY REFERENCE

Passing by reference has caused people much confusion at times. This is a case where using numbers show how it work, or even if it is not.

Not The Same

Passing by value is not the same thing as passing by reference.

Pass by value:

Passing by value will pass a "copy" of the variable value to a function when called. That is what we have been doing up to this point. This is pretty straight forward as said before.

Ie You are using different variables when you call the function to make the flash longer or shorter. We have been doing this with the sos code.

Pass by reference:

When you call the function. The variable will return the address in the memory for the variable.

In other words If you want to return the value assigned in the function. You would call it by reference. Because you will not be changing the variable. You are using the value assigned to the variable. For instance if you are looking for the trigger for an if statement or a switch in case. They are not changing the value, they are testing or printing according to the value stored in the memory address of the variable.

What? You say.

<u>My Number Simulation</u>

Let me put my dummy where my mouth is.

The dummy test is going to be called dummy1PassReference.

This is what I call doing my number simulation.

It is below:

Number Simulation code

```
/* dummy1PassRef .ino */

void setup()

{

    // put your setup code here, to run once:

    Serial.begin(9600);

}

void loop()

{

    // put your main code here, to run repeatedly:

    // Attributes

    int a = 5;

    int b = 15;



    passByValue(a);
```

```
    passByReference(b);



    Serial.println( " ");

    Serial.println( "a returns a copy of the variable: ");

    Serial.println(a);

    Serial.println( "_____");

    Serial.println( " ");

    Serial.println( "b returns the address of the variable:
");

    Serial.println(b);

    Serial.println( "_____");
}
int passByValue(int placeholder1)

{


    placeholder1 = 10;
```

```
}

int passByReference(int& placeholder2)

{

    placeholder2 = 20;

}
```

The syntax for passing by reference is:

<type> <funtionName>(<type>& <variableName>)

{

    <Instructions for funtion>

    <Your instructions would include the code for the

    passed variable you are using. Ie if statement,

    switch and case, etc>

}

You may notice that it looks like a pass by value function. It has one difference. The "&" in the parameters. Do not forget that when passing by reference. If you do you may get the wrong value.

Pass By Reference have one difference.

The "&" in the parameters.

```
int passByReference(int& placeholder2)

{

    placeholder2 = 20;

}
```

Go ahead and verify and upload. Look at the serial monitor.

Do you understand what is happening?

A is returning a copy of the variable assigned and called. That is what we want.

Looking at the pass for B, it is not.

Even though B is assigned to the variable for the passByReference function. It ignores the value assigned to the varialbe. It is using the variable assigned to it's memory address in it's defined function. It will not copy the variable. You would not assign or call it here.

```
// put your main co

// Attributes
int a = 5;
int b = 15;

passByValue(a);
passByReference(b);
```

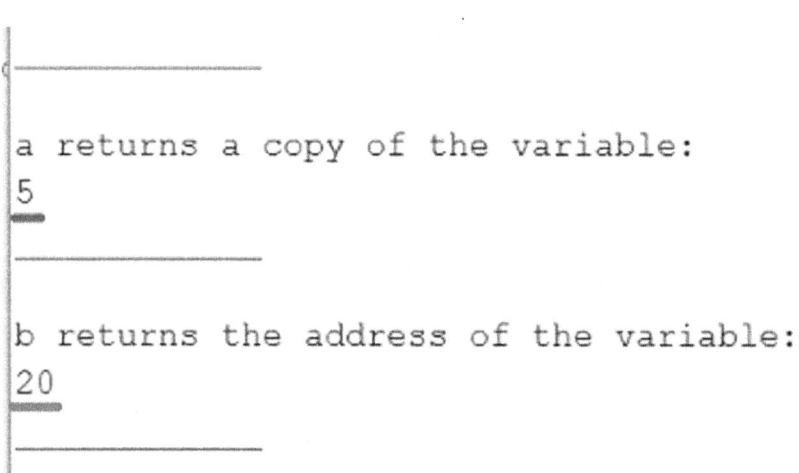

```
a returns a copy of the variable:
5

b returns the address of the variable:
20
```

Take a look at the picture below. It is showing the opposite.

For a it will not use the variable assigned to it in the function(not as shown). The reason for this will be explained later.

For now understand that it can not exist outside its function. So when control pass from the function back to the loop the variable in the function is destroyed.

For b because it is a reference the variable in the function is used. It is not a copy it is the value that is stored in the memory address for the variable.

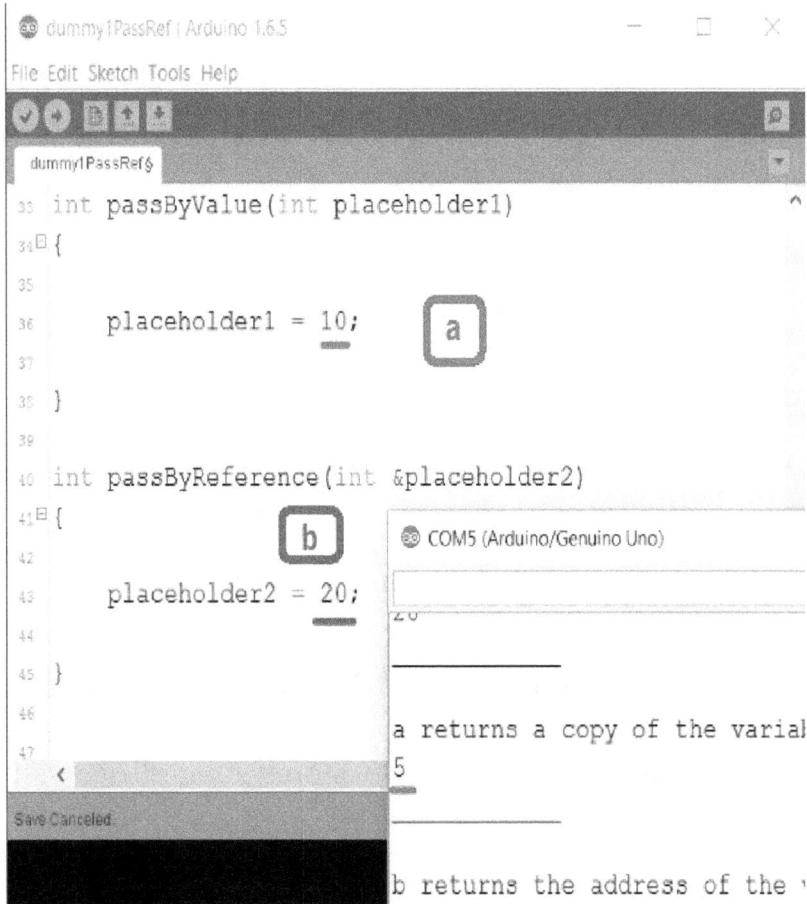

This test shows us that passing by value is what we want for both of our functions.

What was the point of all this.

To show how you may evaluate a process to tell you if it will optimize your code or not.

In this case it will not. In other cases it may.

<u>Getting It Done</u>

Where we left off with our code is below.

```
// morseSos.ino

// Attributes

const int pin13 = 13;      // Stores pin number

const int dwDot = 300;     // Stores 300 millisec value

const int dwDash = 900;    // Stores 900 millisec value

const int dwBetween = 1000;   // Stores 1 sec value
```

```
/*

setup function runs once when you press reset or power
the board

*/

void setup()

{

    // initialize digital pin 13 as an output.

    pinMode(pin13, OUTPUT);

    // Initialize Serial Monitor (more on this later)

    Serial.begin(9600);

}


// the loop function runs over and over again forever

void loop()

{

    // .
```

```
/// S printing S Attribute(variable)

theBlinks(dwDot, pin13, dwBetween);



// -

/// O printing O Attribute(variable)



theBlinks(dwDash, pin13, dwBetween);

// .

/// S printing S Attribute(variable)



theBlinks(dwDot, pin13, dwBetween);



}

// Behaviors
```

```
int theBlinks(int passing1, int passing2, int passing3)

{


    for (int e = 0; e < 3; e++)

    {


        digitalWrite(pin13, HIGH);

        delay(passing1);      // wait variables
milliseconds

        // turn the LED on ( LOW is the voltage level)

        digitalWrite(pin13, LOW);

        delay(passing1);      // wait variables
milliseconds

    }


    // delay 1 seconds between letters
```

```
    delay(passing3);          // wait for a second

    printSos(passing1);

}


int printSos(int passing1)

{


    switch(passing1)

    {

        case 300:  Serial.print(" . . . ");

                   Serial.println("S");

                   break;

        case 900:  Serial.print(" - - - ");

                   Serial.println("O");

                   break;

        default:   Serial.println("Error: Value Not
```

```
Found");

                break;

    }


    return passing1;



}
```

We could set it up to pass by reference but that would be going in the wrong direction. We will keep it the way it is.

Summary:

Whether you use passing by reference or not. It is a good idea to know the in and outs of it.

The main thing that is to be pointed out here is:

You can either pass by value or pass by reference.

They both have advantages and they both work different

ways.

The use of numbers to see how to do something in c++ can help in showing how it works. That is why most examples is given using numbers.

Don't be afraid to try something out. That is the main way you learn when programming.

8 CHAPTER GLOBAL AND LOCAL VARIABLES MEET SCOPE

Did you ever take a look at the black screen at the bottom of the Arduino IDE?

When you don't have an error?

It may say something like:

Sketch uses 2,608 bytes (8%) of program storage space. Maximum is 32,256 bytes.

Global variables use 242 bytes (11%) of dynamic memory, leaving 1,806 bytes for local variables. Maximum is 2,048 bytes.

C:\Program Files (x86)\Arduino\hardware\tools\avr/bin/avrdude -CC:\Program Files (x86)\Arduino\hardware\tools\avr/etc/avrdude.conf -v -patmega328p -carduino -PCOM5 -b115200 -D -Uflash:w:C:\Users\yourName\AppData\Local\Temp\build3088591171746257672.tmp/B21BlinkSosGlobalAndLocalVariables1a.cpp.hex:i

It is giving you some good information. Some is good for setting up an ide if you get to that point. It gives your port number and location of your temp files.

What I want to focus on is the Global and Local variables for this chapter.

They can cause you a world of head ache that can be avoided

Global variables aren't really the problem.

Some folks only use Global variables to keep problems down. That isn't really a good practice though.

Knowing the difference between them can help you when you get the dreaded error.

```
B21BlinkSosGlobalAndLocalVariables1a.ino: In
B21BlinkSosGlobalAndLocalVariables1a:51: erro
'dwDots' was not declared in this scope
```

You will get an error like this from time to time, knowing how to correct it is what's important. You can get this error for many reasons.

A bit of explaining may help you.

Scope

Scope is the reach of a variable. How far it extends in the program. If you get an error like above, it means the variable does not reach, or does not exist. To correct the error the variable would need to be brought into scope.

```
1 // morseSos.ino
2
3 // Attributes
4 const int pin13 = 13;            // Sto
5 const int dwDot = 300;      // Stores
6 const int dwDash = 900;     // Stores
7 const int dwBetween = 1000;    // St
8
9 // setup function runs once when yo
10 void setup()
   <
```

To make things simple for you, we have only been using global variables up to now.

Global Variables

The global variables in the sos sketch have been declared and defined at the top of the sketch over the setup.

They are:

1. Declared before the functions.

2. Compiled before setup and loop

3. Remains in scope for the lifetime of the program.

4. They have Global scope aka program scope.

Global variables are variables that are used by more than one function.

There are ways to get around using them. We have not got to that yet. It will be in upcoming books.

One reason you may get an error for a global variable is you mispelled a variable. Make sure your variables are spelled correctly.

Since global variables have program scope(global scope) . You should not get many errors from them.

So here is your task for now:

Examine the sos code and decide which ones should be global. The solution will be given when we check our code.

Local Variables

Local variables have not been used in the sos code yet, except for the for loop used.

1. They are declared and/or defined inside the function brackets.

2. They are created when the function is ran.

3. Once the function ends the value stored in local variable is destroyed. It is no longer in scope. The variable no longer exist.

4. If you use the same name for that variable in any other function it is a new variable, it has no link to other functions.

5. A local variable has Function scope.

This may be where you get scope errors the most.

When passing variables by value, they can not be defined in the function. Because local variables only exist while in there function. They are destroyed when the function ends. When you call the function the variable value you are trying to pass is destroyed already. Here is where you will get an error.

That is why you can only pass by reference if you are defining the variable in the function. Because it uses the variables memory address.

Again there are more advanced ways to deal with this. For now if you can understand that you are good.

Optimizing The Code

We have four variables to work with here.

Pin13 will be a global variable so it will stay where it is.

It is used in the setup loop and theblinks functions.So best to keep it global for now.

Look at your code again.

For dwBetween, it is only in theBlink. We will add it as a constant in that function. That means that it wil be a local variable to the function theBlink.

Since it is we would delete passing3 out of the function parameters for theBlink.

We now have:

```
// morseSos.ino

// Attributes

const int pin13 = 13;      // Stores pin number

const int dwDot = 300;    // Stores 300 millisec value

const int dwDash = 900;   // Stores 900 millisec value


/*

setup function runs once when you press reset or power the board

*/

void setup()

{
```

```
    // initialize digital pin 13 as an output.

    pinMode(pin13, OUTPUT);

    // Initialize Serial Monitor (more on this later)

    Serial.begin(9600);

}


// the loop function runs over and over again forever

void loop()

{

    // .

    /// S printing S Attribute(variable)

    theBlinks(dwDot);

    // -

    /// O printing O Attribute(variable)

    theBlinks(dwDash);
```

```
//.

/// S printing S Attribute(variable)

theBlinks(dwDot);



}



// Behaviors



int theBlinks(int passing1)

{



    const int dwBetween = 1000;

    // Stores 1 second value



    for (int e = 0; e < 3; e++)
```

```
    {

        digitalWrite(pin13, HIGH);

        delay(passing1);

        digitalWrite(pin13, LOW);

        delay(passing1);

    }

    // delay 1 seconds between letters

    delay(dwBetween);


    printSos(passing1);

    return passing1;

}


int printSos(int passing1)

{

    switch(passing1)
```

```
{

    case 300:   Serial.print(" . . . ");

                Serial.println("S");

                break;

    case 900:   Serial.print(" - - - ");

                Serial.println("O");

                break;

    default:    Serial.println("Error: Value Not
Found");

                break;

    }

    return passing1;

}
```

We are left with the variables dwDot and dwDash, which are the most usable variable we created.

They are only used in the loop function. From there they are passed into theBlink function.

They can be moved to the top of the loop function. Now they will be a local variable to loop.

Verify, Upload, and check the serial monitor.

Everything should work just like it did before.

Checking The Sketch

```
// morseSos.ino

// Attributes

const int pin13 = 13;      // Stores pin number

/*

setup function runs once when you press reset or power
the board.

*/

void setup()

{
```

```
// initialize digital pin 13 as an output.

pinMode(pin13, OUTPUT);

// Initialize Serial Monitor (more on this later)

Serial.begin(9600);

}


// the loop function runs over and over again forever

void loop()

{

    const int dwDot = 300;

    // Stores 300 millisecond value

    const int dwDash = 900;

    // Stores 900 millisecond value


    // .
```

```
    /// S printing S Attribute(variable)


    theBlinks(dwDot);

    // -

    /// O printing O Attribute(variable)

    theBlinks(dwDash);

    // .

    /// S printing S Attribute(variable)

    theBlinks(dwDot);
}


// Behaviors


int theBlinks(int passing1)

{
```

```
    const int dwBetween = 1000;

    // Stores 1 second value


    for (int e = 0; e < 3; e++)

    {

        digitalWrite(pin13, HIGH);

        delay(passing1);

        digitalWrite(pin13, LOW);

        delay(passing1);

    }

    // delay 1 seconds between letters

    delay(dwBetween);              // wait for a second

    printSos(passing1);

    return passing1;

}

int printSos(int passing1)
```

```
{


    switch(passing1)

    {

        case 300:   Serial.print(" . . . ");

                    Serial.println("S");

                    break;

        case 900:   Serial.print(" - - - ");

                    Serial.println("O");

                    break;

        default:    Serial.println("Error: Value Not
Found");

                    break;

    }

    return passing1;

}
```

Summary:

This should clear up the difference between Global variables(program scope) and Local variable(Function Scope).

It may not seem like much but it can help when dealing with scope errors.

Also, remember some scope errors is because a variable may be mispelled. When you correct the spelling it will bring it back into scope..

Above arrays have been mentioned a couple of times. Finally, arrays in chapter 9. That is the next chapter. Have your grapes ready. It is going to be a lot of work(not really), but very beneficial to your programming.

9 CHAPTER ARRAYS

Arrays are like variables. They store values of the same type. Instead of just one value, you can have many.

I think of arrays as Super variables, but that is just me.

The spaces it stores it's data in is called elements.

Just like variables the elements is stored in memory in series.

Below there are 5 elements. Like many things in programming the elements start at 0 not 1.

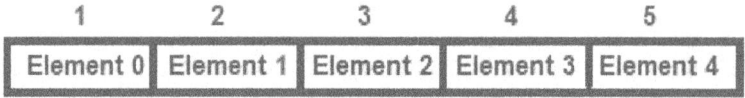

These elements can be accessed by indexing the whole array, or individual elements.

This make it a very good tool for your optimization toolbox.

The program we have been working on is for morse code. So far we have only been using:

s, o, ., and -.

Morse code deals with the whole alphabet and numbers.

Using variables for each letter and number would be valid, You can do it. It would not be optimized and that is what we strive for.

Arrays can solve the problem of optimization for us.

Arrays can be simple or they can be complex.

We are going to do some simple arrays.

Using Arrays With The Functions

Take a deep breath...and here we go!

We are going to add an array for both S and O.

That means we are going to be putting them in two functions.

There will also be a function called beOnOff.

```
int beOnOff(int int1)
{

    digitalWrite(pin13, HIGH);
    delay(int1);
    digitalWrite(pin13, LOW);
    delay(int1);

    return int1;

}
```

The digitalWrite code from theBlink function is being moved to the function beOnOff. This way both arrays will be able to use it. This function is being added at the bottom of the sketch.

With the digitalWrite code moved you are left with.

<table>
<tr><td>Code added at the bottom of the sketch</td></tr>
</table>

```
int theBlinks(int passing1)

{

    const int dwBetween = 1000;

    // Stores 1 second value

    for (int e = 0; e < 3; e++)

    {



    }
```

```
    // delay 1 seconds between letters

    delay(dwBetween);          // wait for a second

    printSos(passing1);

    return passing1;

}
```

We are not going to need the function theBlinks anymore. So lets change the name of the function to:

For the S array: arrForS

Delete passing1 out of all parameters for arrForS function. Delete return passing1.

Do not change the function printSos. It is fine like it is.

You should now have.

```
int arrForS()

{

    const int dwBetween = 1000;

    // Stores 1 second value

    for (int e = 0; e < 3; e++)

    {




    }




    // delay 1 seconds between letters

    delay(dwBetween);          // wait for a second

    printSos();

}
```

Sometimes you will want to update your code that mean

changing it so why not leave the information we need in the function and just rename it.

We will be adding our first array above the for loop.

The syntax is:

// <type> <name> [<size>] = {int1, int2, etc];

We are adding the values for the S blinks.

```
const int dwBetween = 1000;    // Stores 1 second va

int durations[3] =
{

300, 300, 300                  // Array defining delays

};

    for (int e = 0; e < 3; e++)
```

The syntax is pretty straight forward. Notice after the last 300 there is no semicolon or comma. It is supposed to be that way.

Also take note after the closing culy bracket of the array there is a semi colon. That is required. If you don't add it you will get an error.

This is a small very basic array. The values can be for any letter in the alphabet.

Now let's use the array. Because this array is going to allow us to use the beOnOff function and the printSos function.

Backtracking to loops.

Arrays go hand in hand with for loops.

Lucky us. The for loop is already set up from the previous code.

It need to loop through the whole array.

We are going to call the beOnOff function. The array is going to be passed to it's parameters in the loop as shown below.

```
int durations[3] =
{

300, 300, 300                    // Array

};

    for (int e = 0; e < 3; e++)
    {

        beOnOff(durations[e]);

    }
```

Here in the for loop we are calling the beOnOff function. It's parameters is taking the array we created. In the array brackets we are indexing the variable for the for loop.

It is going to loop through the array for the 3 elements.

This is why the for loop is the loop you would use with arrays.

Ok, now let's keep it Rolling and get it printed.

Below is the lines of code we are left with after the for loop in this function.

Code added after the for loop

// delay 1 seconds between letters
 	
delay(dwBetween);   // wait for a second
 	
printSos();
 	
}

For the printSos() function we want a trigger value for the switch and case it holds. Well that is simple. For this one it would be 300 for the delay.

In this case all we need to do is pass it that value.

Just like we passed the whole array in the for loop. We can also pass one element(value) in the printSos parameters.

```
     beOnOff(durations[e]);

   }

   // delay 1 seconds between 1
   delay(dwBetween);

   printSos(durations[0]);

 }

 int printSos(int passing1)
 {
```

The array is then passed to printSos in it's parameters. Just like it was done for beOnOff. But, instead of adding the for loops variable. We add the best element(value) to be the trigger for the switch and case.

Well it was pretty simple since all 3 elements are 300. We could have used 0, 1, or 2.

Now we can call arrForS in the loop and verify and compile.

What we had in the loop before is no good anymore so just delete it.

You code should run nicely for S.

Checking The Sketch

```
// morseSos.ino

// Attributes

const int pin13 = 13;      // Stores pin number



/*

setup function runs once when you press reset or power
the board

*/

void setup()

{

    // initialize digital pin 13 as an output.

    pinMode(pin13, OUTPUT);



    // Initialize Serial Monitor (more on this later)
```

```
    Serial.begin(9600);

}



// the loop function runs over and over again forever

void loop()

{

    // .

    /// S printing S Attribute(variable)

    arrForS();

    // -

    /// O printing O Attribute(variable)

    // .

    /// S printing S Attribute(variable)

}
```

```
// Behaviors

int arrForS()

{

    const int dwBetween = 1000;

    // Stores 1 second value


    int durations[3] =

    {

        300, 300, 300          // Array defining delays

    };


    for (int e = 0; e < 3; e++)

    {

        beOnOff(durations[e]);

    }
```

```
    // delay 1 seconds between letters

    delay(dwBetween);              // wait for a second

    printSos(durations[0]);

}


int printSos(int passing1)

{

    switch(passing1)

    {

    case 300:  Serial.print(" . . . ");

               Serial.println("S");

               break;

    case 900:  Serial.print(" - - - ");

               Serial.println("O");
```

```
                break;

    default:    Serial.println("Error: Value Not Found");

                break;

    }

    return passing1;

}


int beOnOff(int int1)

{

    digitalWrite(pin13, HIGH);

    delay(int1);

    digitalWrite(pin13, LOW);

    delay(int1);

    return int1;

}
```

Did this code optimize what we had already? It all depends on the way you look at it.

Go ahead and try to do O. As always work your way through it. We walked through S and that should give you a roadmap.

The Full Code

```
// morseSos.ino

// Attributes

const int pin13 = 13;      // Stores pin number



/*

setup function runs once when you press reset or power
the board

*/

void setup()

{
```

```
    // initialize digital pin 13 as an output.

    pinMode(pin13, OUTPUT);



    // Initialize Serial Monitor (more on this later)

    Serial.begin(9600);
}



// the loop function runs over and over again forever

void loop()

{

    // .

    /// S printing S Attribute(variable)

    arrForS();

    // -

    /// O printing O Attribute(variable)
```

```
    arrForO();


    // .

    /// S printing S Attribute(variable)

    arrForS();
}


// Behaviors


int arrForS()

{

    const int dwBetween = 1000;

    // Stores 1 second value


    int durations[3] =
```

```
{

    300, 300, 300          // Array defining delays

};


for (int e = 0; e < 3; e++)

{

    beOnOff(durations[e]);

}


// delay 1 seconds between letters

delay(dwBetween);

// wait for a second

printSos(durations[0]);

}
```

```
int arrForO()

{


    const int dwBetween = 1000;

    // Stores 1 second value


    int durations[3] =

    {

        900, 900, 900          // Array defining delays

    };


    for (int e = 0; e < 3; e++)

    {

        beOnOff(durations[e]);
```

```
    }


    // delay 1 seconds between letters

    delay(dwBetween);            // wait for a second


    printSos(durations[0]);


}


int printSos(int passing1)

{

    switch(passing1)

    {

        case 300:  Serial.print(" . . . ");

                   Serial.println("S");

                   break;
```

```
        case 900:   Serial.print(" - - - ");

                    Serial.println("O");

                    break;

        default:    Serial.println("Error: Value Not
Found");

                    break;

    }

    return passing1;

}


int beOnOff(int int1)

{

    digitalWrite(pin13, HIGH);

    delay(int1);

    digitalWrite(pin13, LOW);

    delay(int1);
```

```
    return int1;

}
```

Summary:

If you like variables then you should really like arrays.

Arrays can even hold the alphabet. Even better it can hold the morse codes for the alphabet.

Arrays are for storing a lot of values instead of just one as you do with a variable.

Again they can get very complex. Or they can be quite simple. Simple is better.

This was a very basic example of arrays.

As with all programming, this is something you want to dig in and learn. Practice is the key.

This code have a lot of room for improvement. People

get very creative with their style of code.

Want to show off your code, Become my friend of facebook: Arduino Uno – Ebonygeek45.

10 CHAPTER IN CLOSING

Two boxes of cookies were harmed in the making of this book. Love those chocolate chip pecan cookies.

Congratulations, That is the basics of programming your uno. Some of the things I shared in these books took me a long time to learn. I feel that sharing knowledge is the best thing you can do in life.

The very minimal electronic components was chosen to show results for the coding in this book. That was done intentionally.

You may want to keep these books on hand for reference. Sometimes you need to refresh your knowledge.

In no way is this all you need to learn. This is the basic level c++ side of it. There is also:

- the Arduino side of it,

- the electronics side of it.

- The more advanced c++ side of it.

- Classes

- Libraries

- Pointers

- Robotics

- Alarm Systems

You name it.

But this should give you a sturdy foundation hopefully.

I would request that you feed the muse so to speak. These two books are to help others if they need it. I do need to know if they are well received or if you like them and would want more. If so I will keep writing them to get the knowledge out there.

ebonynerd45@gmail.com

It has been a treat going further on this journey with you.

Happy tinkering, Happy coding.

Today is a good day to learn something.

Ebonygeek45

The End

ABOUT THE AUTHOR

Things may be looking up. Every extra penny still goes to getting the things I need to share my knowledge. I am self taught and the internet is my classroom where I learn. My videos are on YouTube under Ebonygeek45. If I can and do. Then you can to.

I thoroughly enjoyed the creation of Beginners Edition 1, and could not wait to start on this one. More soon to come. Enjoy.

www.ingramcontent.com/pod-product-compliance
Lightning Source LLC
Chambersburg PA
CBHW051702170526
45167CB00002B/501